vjjnf
523.41 DEVER    C0-DYA-893

Devera, Czeena, author
Mercury
33410016576698    05/29/20

Valparaiso Public Library
103 Jefferson Street
Valparaiso, IN 46383

my guide to the planets

# Mercury

# CHERRY LAKE PRESS

Published in the United States of America by Cherry Lake Publishing
Ann Arbor, Michigan
www.cherrylakepublishing.com

Reading Adviser: Marla Conn, MS, Ed, Literacy specialist, Read-Ability, Inc.
Book Designer: Jennifer Wahi
Illustrator: Jeff Bane

Photo Credits: ©Vadim Sadovski/Shutterstock, 5, 21; ©Tom Wang/Shutterstock, 7; ©3Dsculptor/Shutterstock, 9; ©Baranov E/Shutterstock, 11; ©Mopic/Shutterstock, 13; ©NASA images/Shutterstock, 15, 23; ©PIA18215/NASA, 17; ©Jurik Peter/Shutterstock, 19; Cover, 2-3, 6, 8, 22, Jeff Bane; Various vector images throughout courtesy of Shutterstock.com

Copyright ©2020 by Cherry Lake Publishing
All rights reserved. No part of this book may be reproduced or utilized in any form or by any means without written permission from the publisher.

Library of Congress Cataloging-in-Publication Data

Names: Devera, Czeena, author. | Bane, Jeff, 1957- illustrator. | Devera, Czeena. My guide to the planets.
Title: Mercury / by Czeena Devera ; illustrated by Jeff Bane.
Description: Ann Arbor, Michigan : Cherry Lake Publishing, [2020] | Series: My guide to the planets | Includes index. | Audience: K-1.
Identifiers: LCCN 2019032932 (print) | LCCN 2019032933 (ebook) | ISBN 9781534158863 (hardcover) | ISBN 9781534161160 (paperback) | ISBN 9781534160019 (pdf) | ISBN 9781534162310 (ebook)
Subjects: LCSH: Mercury (Planet)--Juvenile literature.
Classification: LCC QB611 .D48 2020  (print) | LCC QB611  (ebook) | DDC 523.41--dc23
LC record available at https://lccn.loc.gov/2019032932
LC ebook record available at https://lccn.loc.gov/2019032933

Printed in the United States of America
Corporate Graphics

## table of contents

About Mercury . . . . . . . . . . . . . . . . . 4

Glossary . . . . . . . . . . . . . . . . . . . . 24

Index . . . . . . . . . . . . . . . . . . . . . . 24

**About the author:** Czeena Devera grew up in the red-hot heat of Arizona surrounded by books. Her childhood bedroom had built-in bookshelves that were always full. She now lives in Michigan with an even bigger library of books.

**About the illustrator:** Jeff Bane and his two business partners own a studio along the American River in Folsom, California, home of the 1849 Gold Rush. When Jeff's not sketching or illustrating for clients, he's either swimming or kayaking in the river to relax.

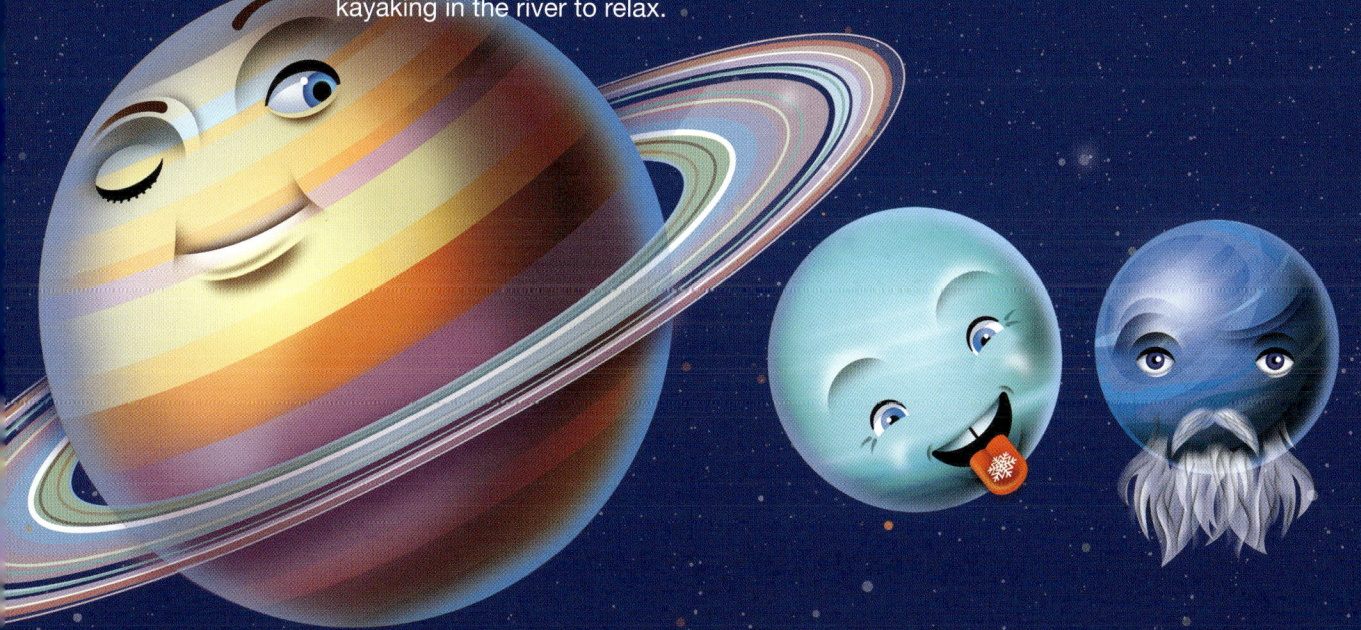

## About Mercury

I'm Mercury. I am the closest planet to the Sun.

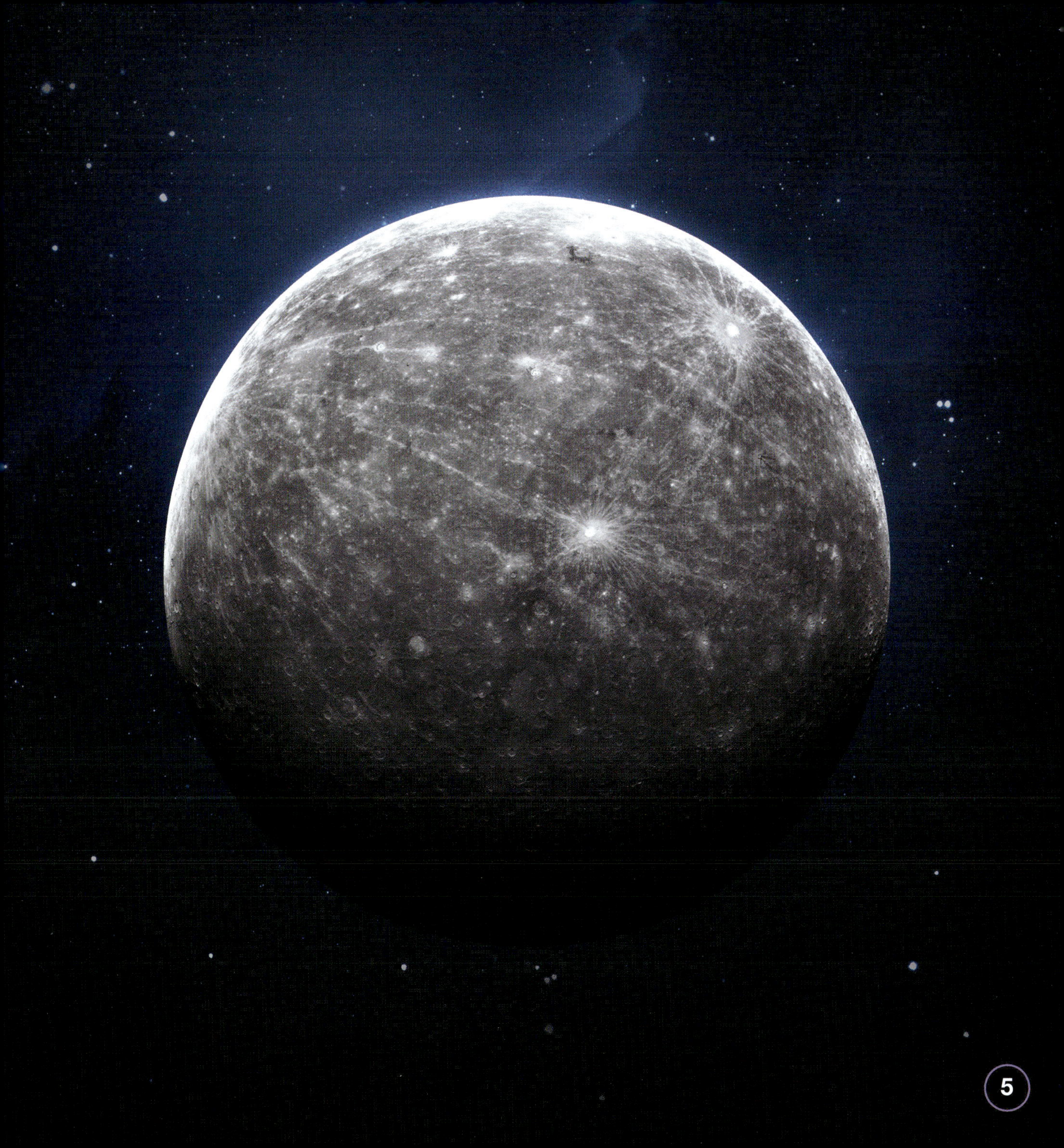

But I'm not the hottest planet.

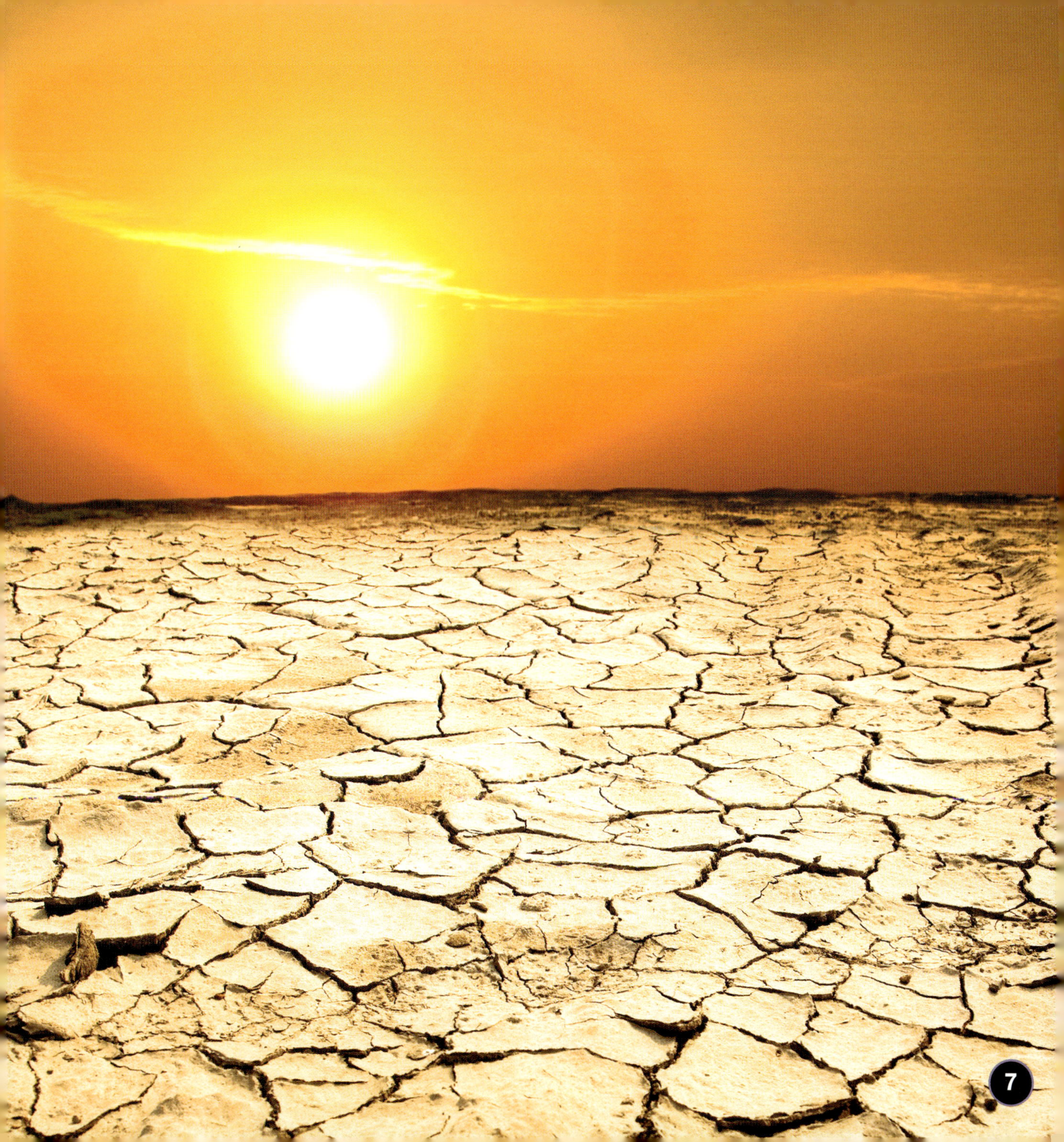

Why do you think that is?

I'm hard to visit. Only two spacecrafts have ever gotten close to me.

9

I am the smallest of all the planets. I am only a little bit larger than Earth's Moon.

11

I **orbit** around the Sun. It only takes me 88 days to complete 1 orbit!

13

My days are long. One entire day for me is 58 days on Earth!

15

I have a pretty rocky surface.
I have many **craters**.

17

I don't have any rings. I don't have any moons either.

19

You might have seen me. I can be seen from Earth every 7 years.

21

What makes me unique?

I am a **unique** planet. **Scientists** are discovering new things about me!

23

## glossary & index

### glossary

**craters** (CRAY-turz) large holes in the ground caused by something falling, like meteorites

**orbit** (OR-bit) to travel in a curved path around something

**scientists** (SYE-uhn-tists) people who study nature and the world we live in

**unique** (yoo-NEEK) the only one of its kind

### index

Earth, 10, 14, 20

craters, 16

moon, 10, 18

orbit, 12

rings, 18

spacecrafts, 8
Sun, 4, 12